KNOW THYSELF.
KNOW THY BRAIN.
KNOW THY GOD.

*STRATEGIES TO REWIRE THE BRAIN WITH
THE WORD OF GOD*

RENÉE D. CHARLES, PH.D.

Copyright © 2022 by Renee D. Charles, Ph.D. All rights reserved.

It is in no way legal to reproduce, duplicate, or transmit any part of this document in any form by any means—electronic, mechanical, photocopy, recording, or otherwise—without prior written permission from the publisher, except as provided by United States of America copyright law.

Unless otherwise noted, all Scripture quotations are taken from the King James Version®. Copyright © 1982 by Thomas Nelson. Used by permission. All rights reserved.

Scripture quotations marked NKJV are taken from the New King James Version®. Copyright © 1982 by Thomas Nelson. Used by permission. All rights reserved.

Scripture quotations marked NIV are taken from the Holy Bible, New International Version®, NIV®. Copyright © 1973, 1978, 1984, 2011 by Biblica, Inc.™ Used by permission of Zondervan. All rights reserved worldwide www.zondervan.com. The "NIV" and "New International Version" are trademarks registered in the United States Patent and Trademark Office by Biblica, Inc.™

Editing by Elite International Publishing

Cover design by: Xee_designs1

Visit the author's website at www.drreneecharles.com Library of Congress Cataloging-in-Publication Data:

An application to register this book for cataloging has been submitted to the Library of Congress.

ISBN: 978-1-7338174-4-8

This book contains the opinions and ideas of its author. The information provided in this book is for educational purposes only and should not interfere with current or future relationships with healthcare providers or legal counsel. Under no circumstances will the publisher be held legally responsible or blamed for any reparation, damages, suffering, or monetary loss due to the information herein, either directly or indirectly.

I have chosen to adopt Bishop T. D. Jake's position to not capitalize "s" in the spelling of the name of satan out of a lack of respect for him.

Printed in the United States of America

PREFACE

Neuroscience studies the brain, the nervous system, function, and how it works when engaged in a task.

Feelings are our primary way of interacting with the world through our five senses (touch, taste, smell, hear and see) and the conscious experience of an emotional reaction.

Physical and internal Trauma affects our emotional experience and brain circuitry, creating scars and deep wounds we often don't recognize. Feelings are not emotions, but sensations that arise in the body activated through chemicals (neurotransmitters and hormones) released by the brain in response to a conscious or subconscious interpretation of a specific trigger.

To discern Trauma, we must fully understand our emotions and our brain.

This book describes from a Biblical perspective how to rewire the brain with the Word of God and apply neuro-strategies to heal Trauma wounds and bruises. Recovery, like salvation, is a restorative process to return the individual to a superior pre-injury state. As you take authority over your thoughts with the Word of God, new spurts of energy will begin to stream throughout the brain, creating new neural pathways and ways of thinking. Over time with consistency and intentionality, the brain will choose behaviors that have been rewarding in the past.

CONTENTS

Preface iv
Introduction vi

KNOW THYSELF 1
KNOW THY BRAIN 3
KNOW THY GOD. 5
A BIBLICAL EXAMPLE OF REWIRING YOUR BRAIN IN FAITH 6
BLUEPRINT FOR REWIRING THE BRAIN 8

Endnotes 13

INTRODUCTION

Childhood neglect, abandonment, rejection, betrayal, and sexual abuse, among other traumatic events, can wreak havoc on our bodies and create strongholds in our minds. Where the mind goes, the body follows.

This writing includes excerpts from my book Remembering the Trauma and Healing it With the Trauma of Change System Model to fast-track strategies to rewire your brain with the Word of God.

Behind every stronghold, a framework of Trauma is a spirit aiming to incapacitate unbelievers and believers alike. The Spirit of Trauma as described in Remembering the Trauma and Healing it With the Trauma of Change System Model strangles people with feelings of shame and guilt about their past and present lives, which hinders their ability to manifest the glory of God—even after accepting Christ as Lord and Savior.

Trauma happens, and when it does, it occurs unannounced and often without warning or permission. "Trauma is an event that overwhelms the central nervous system, altering how we process and recall memories." When our internal resources are inadequate to cope with the external threat, Trauma happens.

Trauma is the Greek Word for "wound" a hurt, or defeat; the term can also mean "to pierce, damage defeat." Sigmund Freud defined Trauma as "a breach of the protective barrier against stimuli leading to feelings of overwhelming helplessness. "Following the initial trauma, a person may be left with intrusive thoughts, memories, flashbacks, and nightmares, triggering the "fight, flight, or freeze response".

CHAPTER 1

KNOW THYSELF

The mind is your soul (mind, will, and intellect); it is metaphysical; it's invisible; it's the spiritual part of humanity extending beyond our physical selves with limitless possibilities. It is the essence of your being, our perceptions, ideas, and beliefs about the world and our lived experiences.

The mind, a type of software that regulates the brain's flow of information, is a stream of nonconscious and conscious (thoughts) when we are awake and nonconscious activity (energy) when we are asleep. The brain takes the shape that the mind rests upon. Thoughts are electrochemical reactions, energy signals (firing neurons) generated in the brain, composed of 100 billion nerve cells (neurons) that transmit impulses through synapses.

Our thoughts create emotions, and emotions give more energy and power to the originating idea. When you think, you will feel; when you think and feel, you will choose or decide, which drives our actions. These three aspects always work together. Everyone processes information with both the rational and emotional parts of the brain. The amygdala (emotional brain) is "the integrative center for emotions, emotional behavior, and motivation; it is wired to detect threats and involve how we make decisions, form memories, and encode negative memory into the hippocampus.

People who have been traumatized hold an implicit memory of traumatic events in their brains and body. You can consciously decide to hack your brain and reroute your thoughts to a higher level. One way to change repetitive negative thought processes is to disallow the counterfeited mental picture to have the first position in your thought process. Over time you will establish a new mental map. A mental map is a first-person perception of one's world. Your new mental map will become automatic when more efficient thoughts are superimposed over old, disabling ideas, new neural pathways are created.

CHAPTER 2

KNOW THY BRAIN.

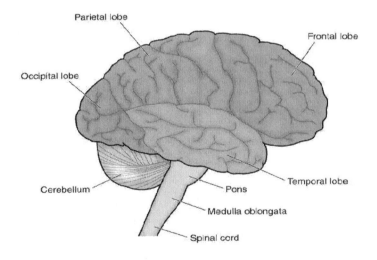

Figure 1: Four Lobes of the Cerebrum

The brain is a three-pound physical organ and a habit-forming machine; a type of hardware moving parts and processing information. The brain has a size and shape comprised of (right and left hemispheres) that we can touch.

The brain like the triune nature of God (Father, Son, and Holy Spirit), is a threefold cord that is not easily broken (Eccl. 4:12). Our brain is the most important and complex organ in our bodies).

From the top down is comprised of three parts (cerebrum, the rational brain or executive state, which is the frontal cortex, which is responsible for problem-solving, memory, language, judgment, impulse control, and reasoning.

The second level is the emotional state, processing emotions such as fear, pleasure, or anger. Emotions drive 80% of the choices Americans make. The third level the brain stem is the survival state, (primal brain) and is responsible for survival, drive, and instinct.

CHAPTER 3

KNOW THY GOD.

There is a war going on in the realm of the spirit. The battle is over your body, soul, and spirit. If you are a believer, your spirit has been redeemed by the blood of the Lamb. Your soul is being redeemed through the renewing of your mind. Your body will be redeemed again when Jesus returns.
Our brain is a goal-achieving machine with the capacity to know and communicate with God through the human Spirit. We can train our minds to retrain our brains to be intentional and decide actions based on the Word of God.

Recovery, like salvation, is a restorative process to return the individual to a superior pre-injury state. As you take authority over your thoughts with the Word of God, new spurts of energy will begin to stream throughout the brain, creating new neural pathways and ways of thinking. Over time with consistency and intentionality, the brain will choose behaviors that have been rewarding in the past.

CHAPTER 4

A BIBLICAL EXAMPLE OF REWIRING YOUR BRAIN IN FAITH

Figure 2:

King David is a perfect example of how you can rewire your brain in faith. The Word of God refers to King David as a man after God's own heart (Acts 13:22).

David experienced childhood abandonment, being rejected by his father. He then suffered persecution and betrayal trauma from King Saul. Furthermore, King David was betrayed by his son Absalom. King David faced one-step transmission trauma when his son Amnon became obsessed with Tamar, his half-sister, and raped her. Moreover, he lost the child conceived due to the sin he committed with Bathsheba.

Although David experienced multiple severe traumas, he stood on the Word of God. David practiced narrative memory as evidenced by his writing of much of the book of Psalms. Many of the psalms exemplified his faith in God and released his emotions.

"I waited patiently for the Lord, and He inclined to me and heard my cry. He brought me up out of the pit of destruction, out of the miry clay, and He set my feet on a rock making my footsteps firm. He put a new song in my mouth, a song of praise to our God; many will see and fear and will trust in the Lord." (Ps. 40:1–3 NASB)

If the Lord had not been my help, my soul would soon have dwelt in the abode of silence. If I should say, "My foot has slipped," Your loving kindness, O Lord, will hold me up. When my anxious thoughts multiply within me, your consolations delight my soul. (Ps. 94:17–19 NASB)

CHAPTER 5

BLUEPRINT FOR REWIRING THE BRAIN

The first step in rewiring your brain is to learn how to regulate intense emotions. Once emotions are controlled, you can rebuild the ability to trust the Word of God and the people God has assigned to you.

1. Trust in the Word and the promises of God.

Realize that your circumstances have nothing to do with what God has promised you. "Let this mind be in you which was also in Christ Jesus" (Phil. 2:5–11). Trust is a part of your brain's default setting. The anterior cingulate cortex (ACC) is the accountant of the brain, organized to assess reward-risk-based decision-making.

If an individual's actions and beliefs are not aligned, it will trigger an alarm in the amygdala. When there is a breach of trust, the brain's conflict detector, the ACC, activates the amygdala. Trust and fear are inversely related. Anxiety activates the amygdala. Trust decreases the activation of the amygdala. "Trust frees the brain to engage in other activities like creativity, planning, and decision-making"

2. Make a quality decision to adopt new, more effective ways to think, feel, and speak to old, disabling beliefs, triggers, and thoughts. Meditate on Romans 12:2:

"Be not conformed to the world but be transformed by renewing your mind." You must be intentional because God is intentional.

3. Write the thoughts you want to become a part of your mental map. Create a script for each day. The act of writing stimulates the limbic system's emotional processing, right brain creativity, and insights Making a "decision" includes identifying and choosing alternatives, creating intentions, and setting goals. All three decision-making parts are part of the same neural circuitry and positively engage the prefrontal cortex by reducing worry and anxiety.

4. Give yourself permission to rewrite positive narratives and embrace new positive emotions. Emotion is not a thought It is a conscious awareness of a feeling. Writing deepens neural pathways for learning. "And the Lord answered me, and said, Write the vision, and make it plain upon tables, that he may run that readeth it" (Hab. 2:2). Emotions originate as sensations in the body reacting to a positive or negative thought. Feelings are influenced by our emotions but are generated from our thoughts. Human emotions are hypothetically the fuel that drives the car. You can't arrive at your point of destiny on an empty fuel tank. Without an authentic emotion, a thought sits in neutral like a car without gas. Conscious and critical thinking will help you to mentally shift and increase the flow of brain connectivity from the left and right brain, preventing it from being hijacked by the amygdala.

5. Permit yourself to engage in positive thoughts: gratitude and forgiveness coordinate thoughts and emotions. Gratitude is a constant state of mental appreciation. The benefits of gratitude start with the dopamine system because feeling grateful activates the brain stem region that produces dopamine (an excitatory neurotransmitter). Gratitude involves the cerebellum, the emotional brain, and the temporal lobe. Gratitude affects your brain at the biological level and boosts dopamine (a happy feeling hormone). Be intentional in forgiving those who misused and hurt you. In prayer, tell God how grateful you are for His presence, favor, and the ability to forgive as He forgives you.

6. Visualize your destination. You must see yourself where you want to be before you arrive there. See yourself delivered, victorious, and healed from the Spirit of Trauma. The process of visualization primes the brain. You can use visualization to create the emotions you want to go through. The stronger the feelings are, the stronger the mental map will be.

When you visualize your intentions, your brain can't tell the difference between what is real or imagined. As you mentally rehearse new affirmations and habits, you strengthen your ability to create them in your life and open portals in your right

The Word of God reveals for as he thinks in his heart, so is he. (Prov 23:7 NKJV).

7. Speak only the truth of God's Word. You are a speaking spirit—your words matter. Your words carry energy that affects the matter in your life. Do not confess those negative thoughts in your mind; rather, speak life to every situation around you.

Words lack meaning without context. Myles Munroe said, "If the context is wrong, the conclusion is also wrong." It's the intention behind the words that convey the vibration.
Your words must align with God's Word to effect change in yourself and your life. "Study to shew thyself approved unto God, a workman that needeth not to be ashamed, rightly dividing the word of truth" (2 Tim. 2:15). Study and meditate on Scriptures about the change you want to create.

You will overcome as you align your confession with God's Word. Our words have the power to destroy and the power to build up (Prov. 12:6). The writer of Proverbs tells us, "The tongue has the power of life and death, and those who love it will eat its fruit" (Prov. 18:21 NIV).
Speak only the truth of God's Word. You are a speaking spirit—your words matter. Your words carry energy that affects the matter in your life. Do not confess those negative thoughts in your mind; rather, speak life to every situation around you.

Just as God has come for our words, satan comes for your words as well. Anytime you open your mouth; you are on trial; words create worlds.

For by thy words thou shalt be justified, and by thy words, thou shalt be condemned (Matt. 12:37).

8. Practice the thoughts you want to predominate your waking day and dreams. New ideas and experiences create new neural pathways, thought processes, and mental associations. Your mind becomes renewed as you establish new patterns and practices. Practice thinking, feeling, visualizing, and acting in alignment with your desired intentions to remove old behaviors and thoughts. When you do this, you give notice to demonic squatters that their time is up by commanding them in the name of Jesus to leave and never return. Repeating new and purposeful actions will become an ingrained and habitual thought process. Be ye transformed by the renewing of your mind.

"And be not conformed to this world: but be ye transformed by the renewing of your mind, that ye may prove what is good, acceptable, and perfect, will of God" (Rom. 12:2).

9. Declare God's Word to override your past. Superimpose the truth of what God says about you over the painful, old negative memories.

"It is the Spirit that quickeneth; the flesh profiteth nothing: the words that I speak unto you, they are spirit, and they are life." (John 6:63)

10. As you meditate and speak the Word of God, allow your heart to release and disallow emotions of anxiety, pain, shame, fear, and guilt.

Introduce yourself to a new feeling of peace, confidence, hope, and love. Reading and memorizing scriptures won't be enough if you don't believe and apply the Word to your life. A brain is a habit-forming machine. Practice and repeat what the Word of God says until it becomes a part of your mental map. Tap into your heavenly language by praying in the Spirit (in tongues; satan cannot discern what you are saying to God as you pray in supernatural tongues). Praying in the Spirit allows the Word of God to have predominance in your mind.

When you apply these strategies and declare the Word over your life, it will become faithful to you. And be not conformed to this world: but be ye transformed by the renewing of your mind, that ye may prove what that good is, and acceptable, and perfect, will of God. Where your mind goes, the body will follow.

ENDNOTES

https://www.psychotherapy.net/interview/bessel-van-der-kolk-trauma

(https://www.etymonline.com/word/trauma, n.d.)

http://docplayer.net/31946608-Healing-the-trauma-body-by-william-smythe-certified-advanced-rolfer.html

https://www.mindfulheartsinstitute.com/complex-trauma-and-ptsd/

https://nba.uth.tmc.edu/neuroscience/m/s4/chapter06.html

(https://www.crisisprevention.com/Blog/how-the-primal-brain-affects-behavior , n.d.)

http://www.hrzone.com/engage/managers/the-neuroscience-of-trust-and-how-it-can-improve-your-engagement-results

https://catnipblog.com/tag/creating-intentions-setting-goals.

https://herheart.org/4-rituals-that-will-make-you-happier/

brain.https://www.usingenglish.com/forum/threads/circuit-switch.229748/